MW01531279

World Book, Inc.
180 North LaSalle Street
Suite 900
Chicago, Illinois 60601
USA

For information about other "True or False?" titles, as
well as other World Book print and digital publications,
please go to www.worldbook.com.

For information about other World Book publications,
call 1-800-WORLDBK (967-5325).

For information about sales to schools and libraries,
call 1-800-975-3250 (United States) or 1-800-837-5365
(Canada).

Library of Congress Cataloging-in-Publication Data for
this volume has been applied for.

True or False?
ISBN: 978-0-7166-4069-1 (set, hc.)

Robots
ISBN: 978-0-7166-4078-3 (hc.)

Also available as:
ISBN: 978-0-7166-4088-2 (e-book)

Printed in the United States of America by CG Book
Printers, North Mankato, Minnesota

1st printing March 2020

Staff

TRUE OR FALSE?

ROBOTS

WORLD BOOK

www.worldbook.com

TRUE OR FALSE?

Robots are all built to look like humans.

4

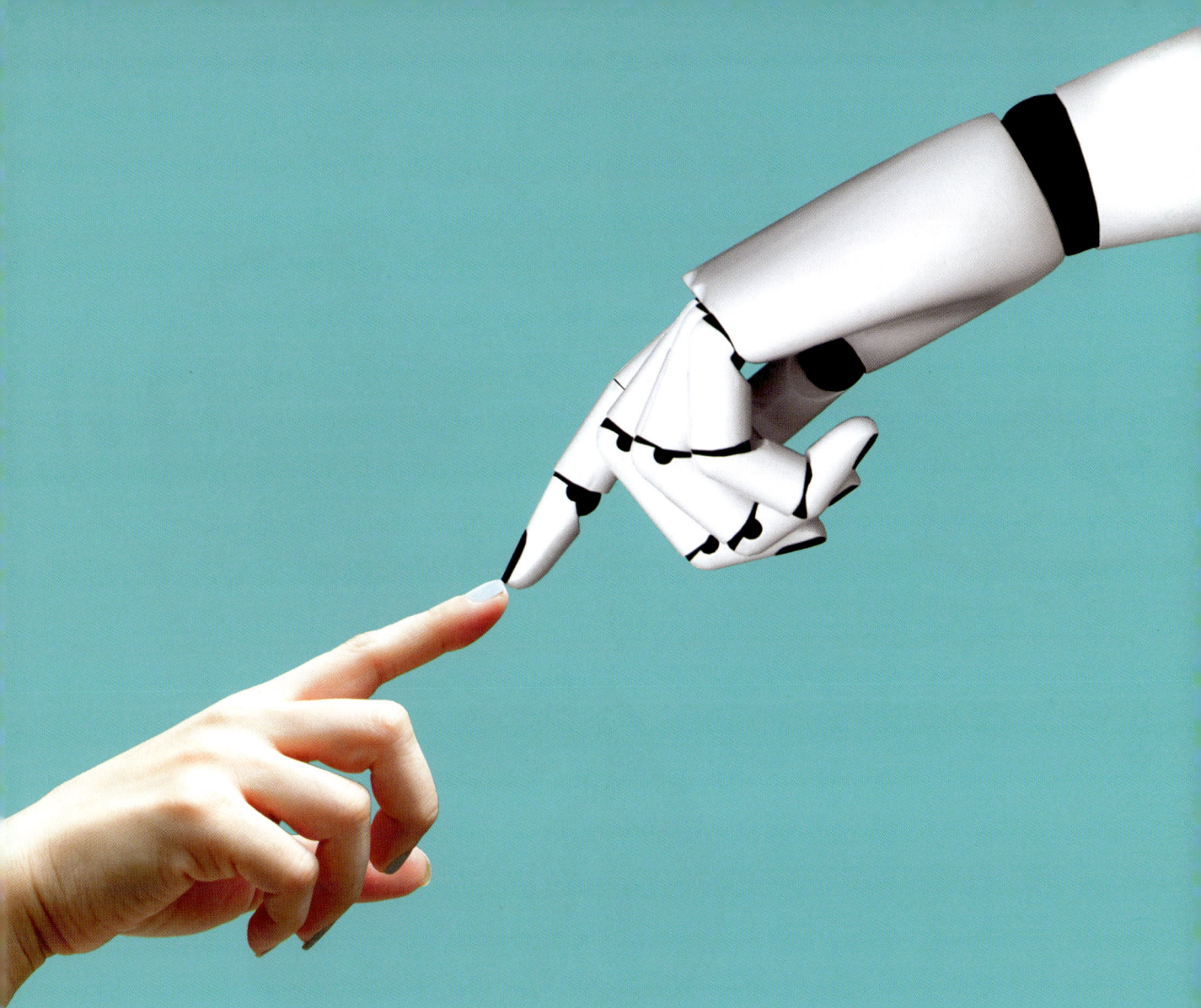

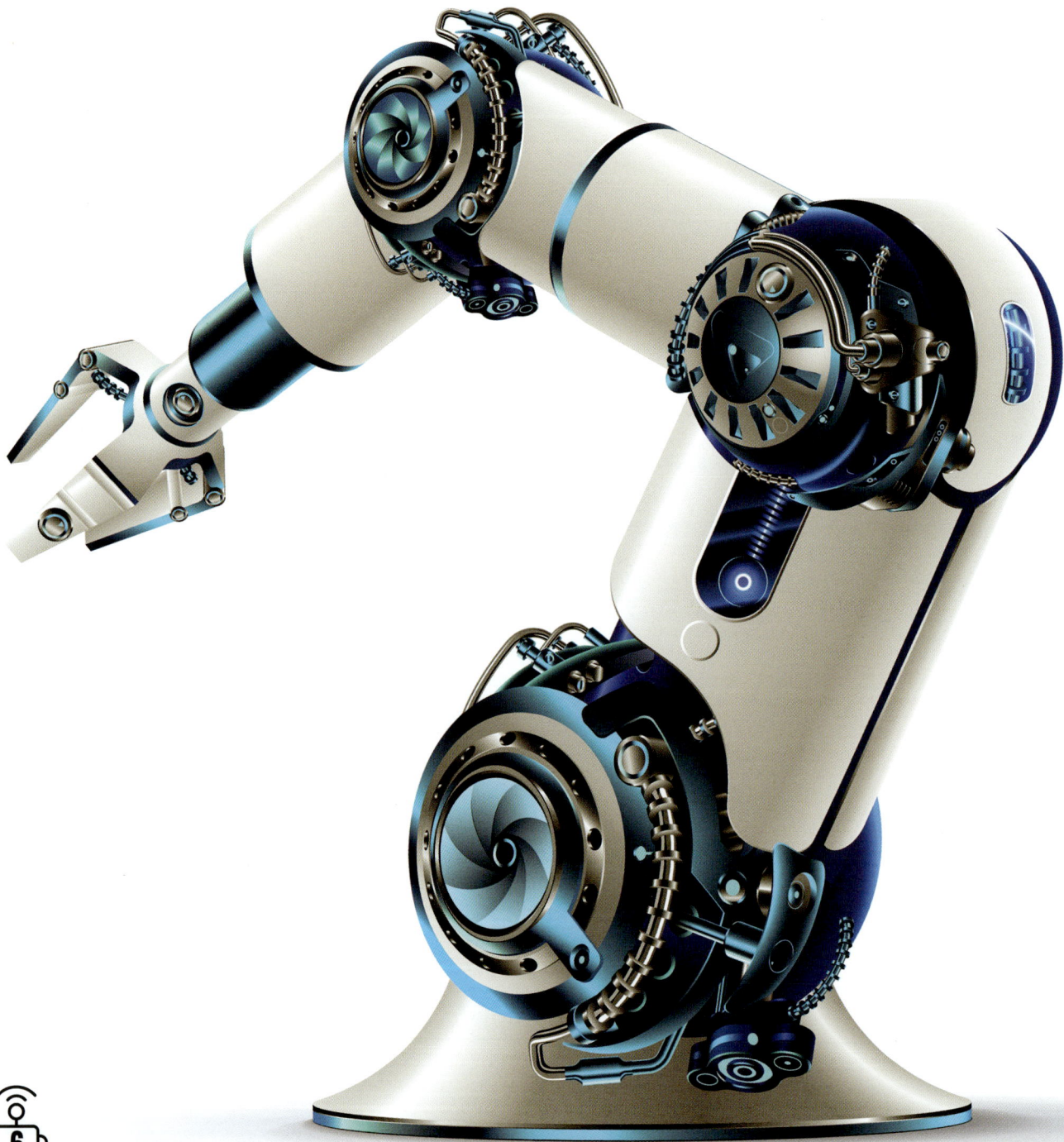

FALSE!

Robots come in all shapes and sizes. A *robot* is any machine that can plan and carry out a series of actions or tasks all by itself.

TRUE OR FALSE?

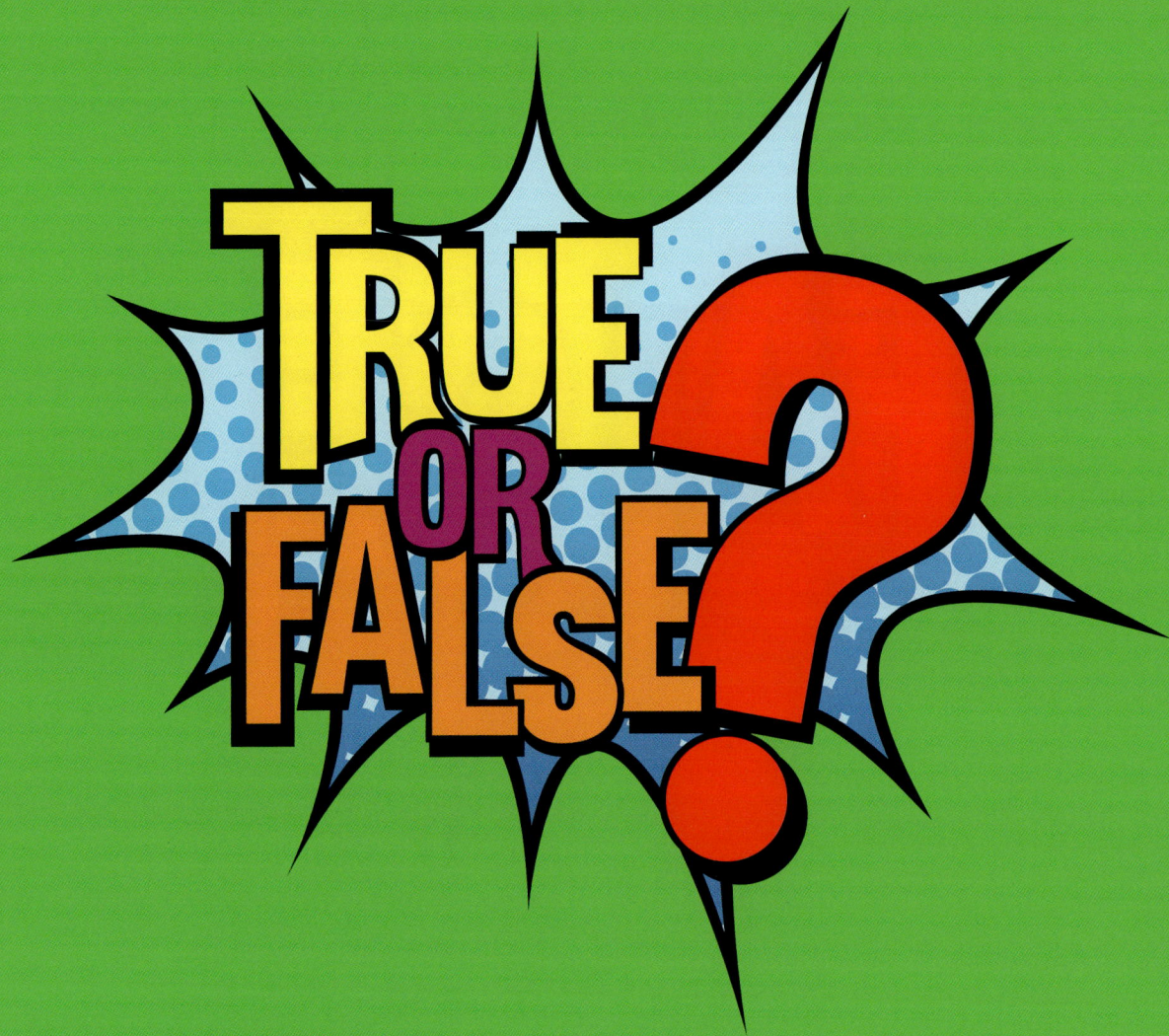

Robots are all around us.

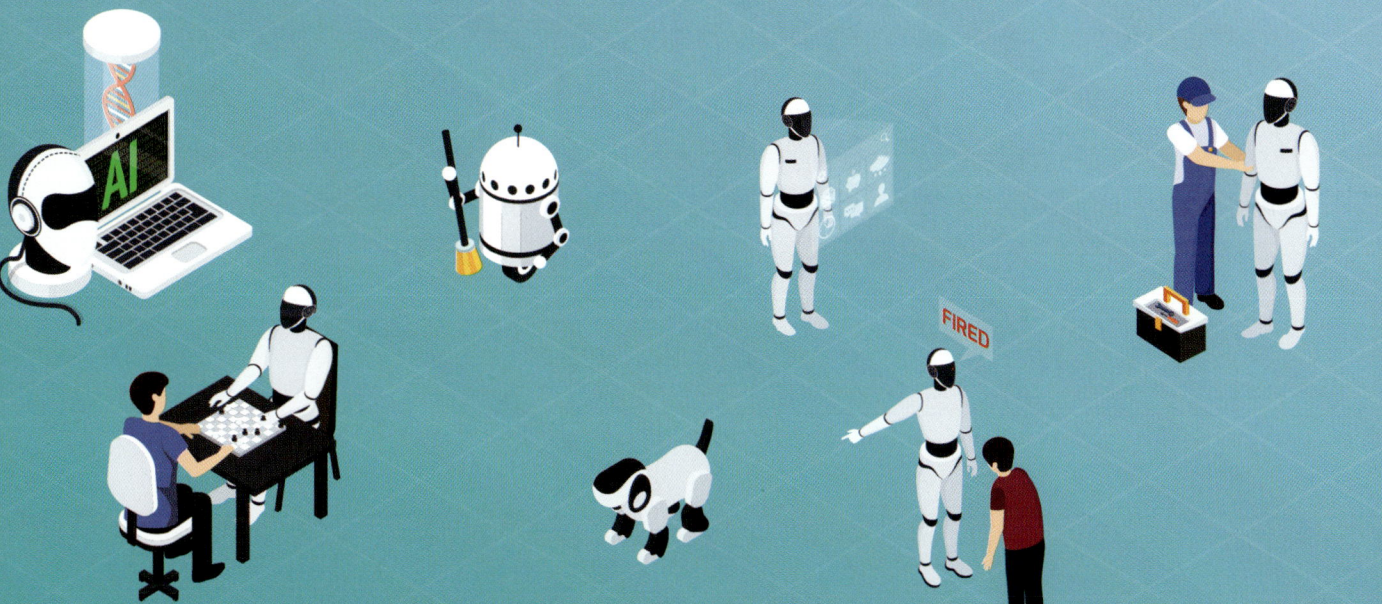

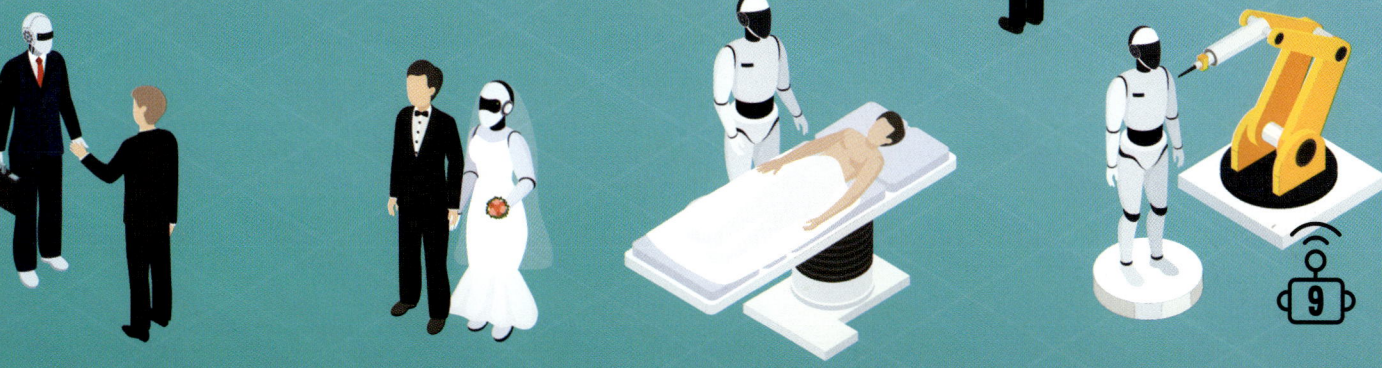

TRUE!

Robots are part of modern life. Robots clean floors, build products used every day, and even perform surgeries!

TRUE OR FALSE?

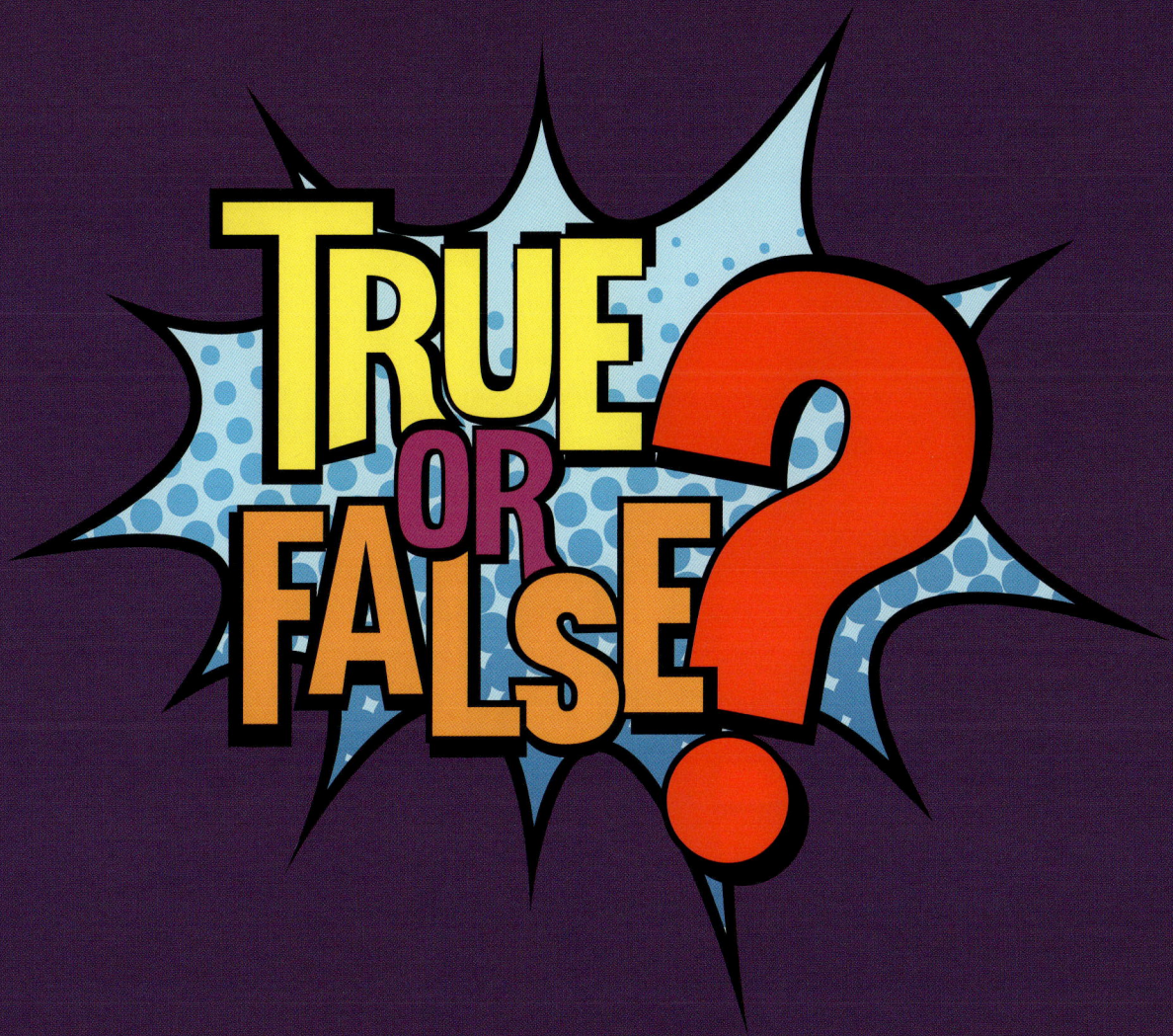

Robots do the coolest jobs.

13

FALSE!

Most robots do work that is really boring, like putting the same two parts together over and over again.

15

TRUE OR FALSE?

Robots make all kinds of products, but they don't work well with food.

17

FALSE!

Robots help make and package many foods sold in grocery stores. They are also branching out into harvesting fruit and flipping burgers.

18

TRUE OR FALSE?

All robots are able to move.

21

TRUE!

A robot must move in some way to interact with the world around it. It might stay in one place, though, like a robotic arm bolted to a factory floor.

SCANNING
IN
PROGRESS

TRUE OR FALSE?

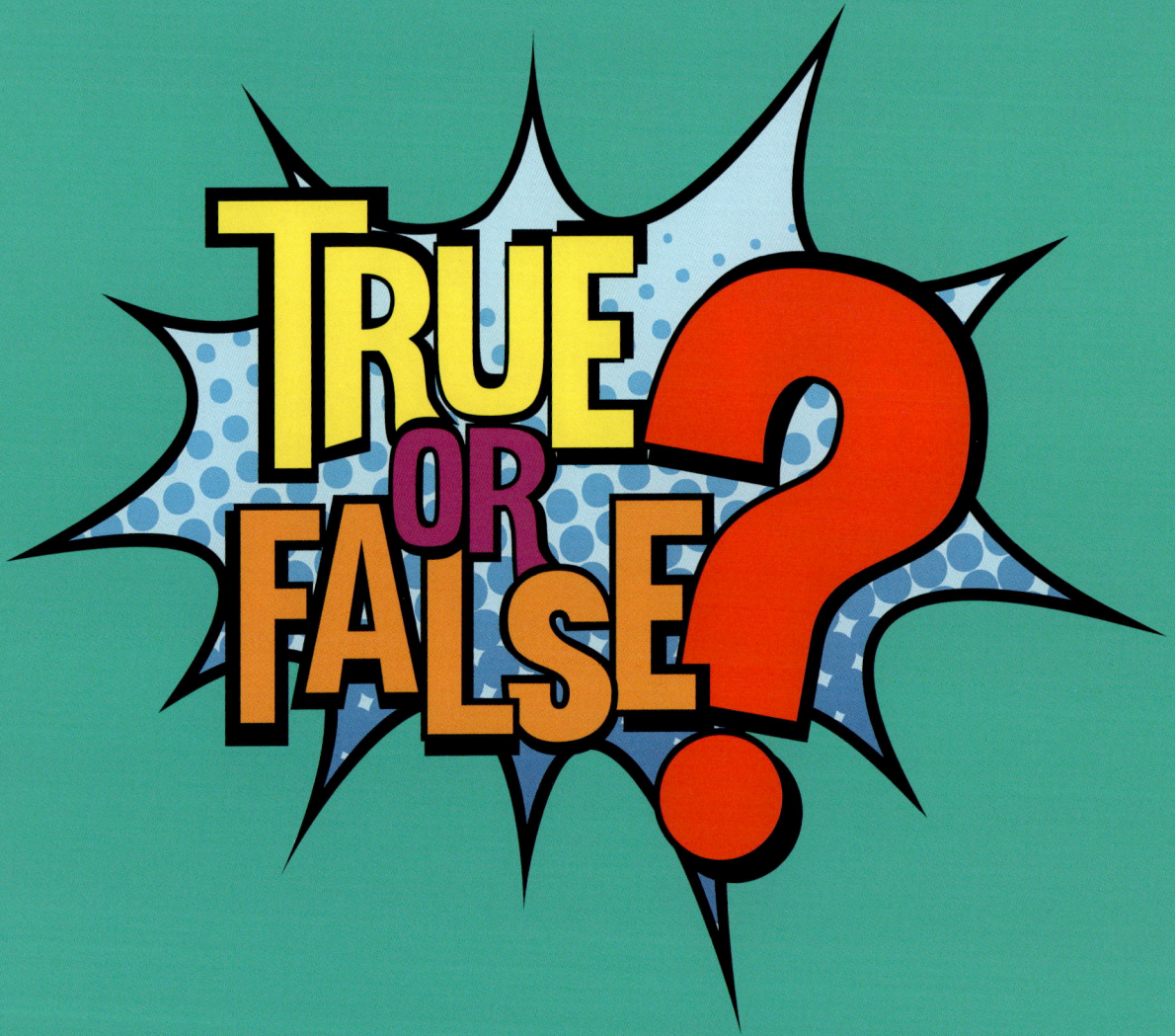

All robots can sense their surroundings.

24

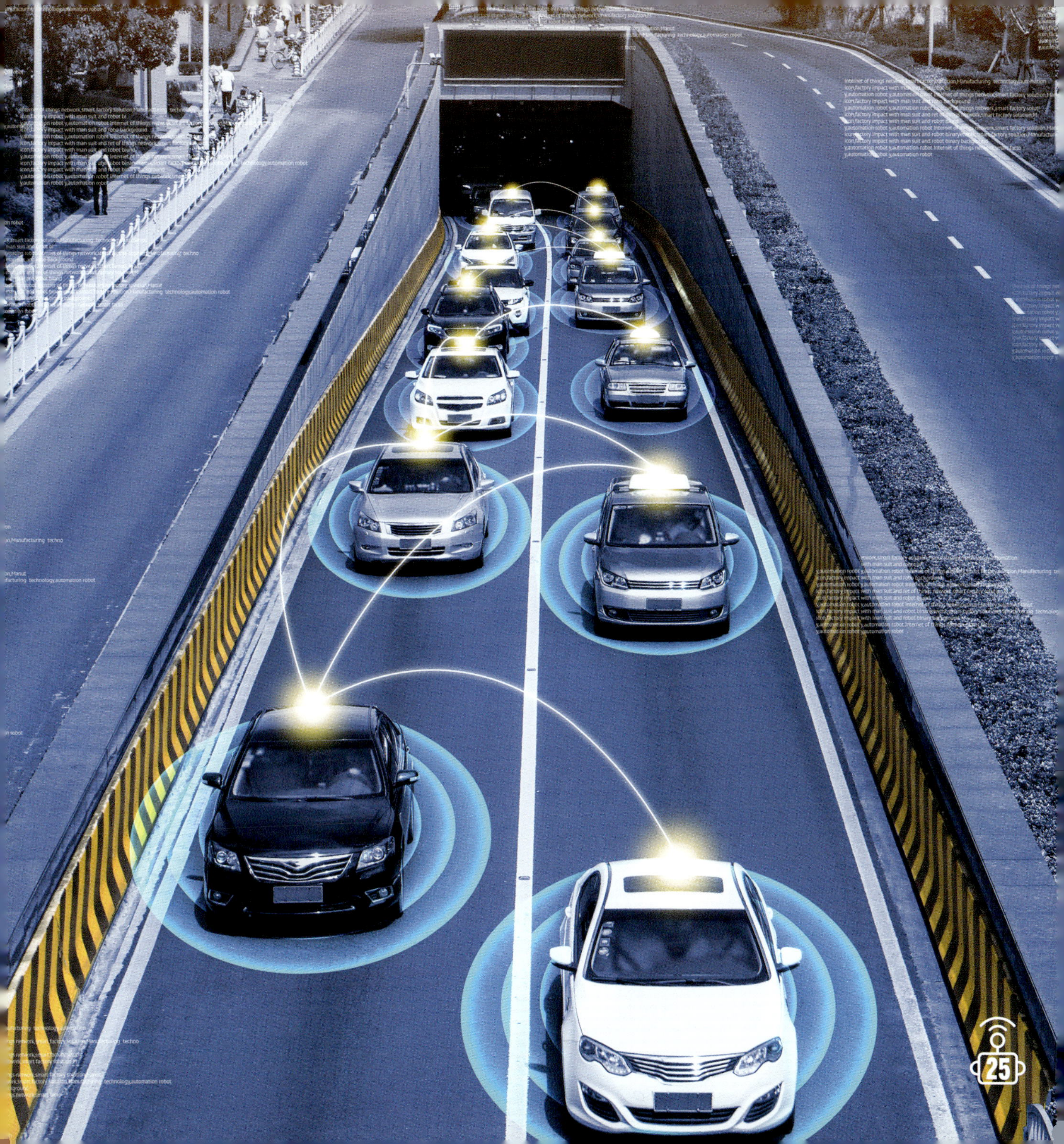

TRUE!

Robots use sensors to get information about what's around them. For example, if a robot senses an object in its way, it might move around it. Without sensors, a robot would be nothing but an ordinary machine.

TRUE OR FALSE?

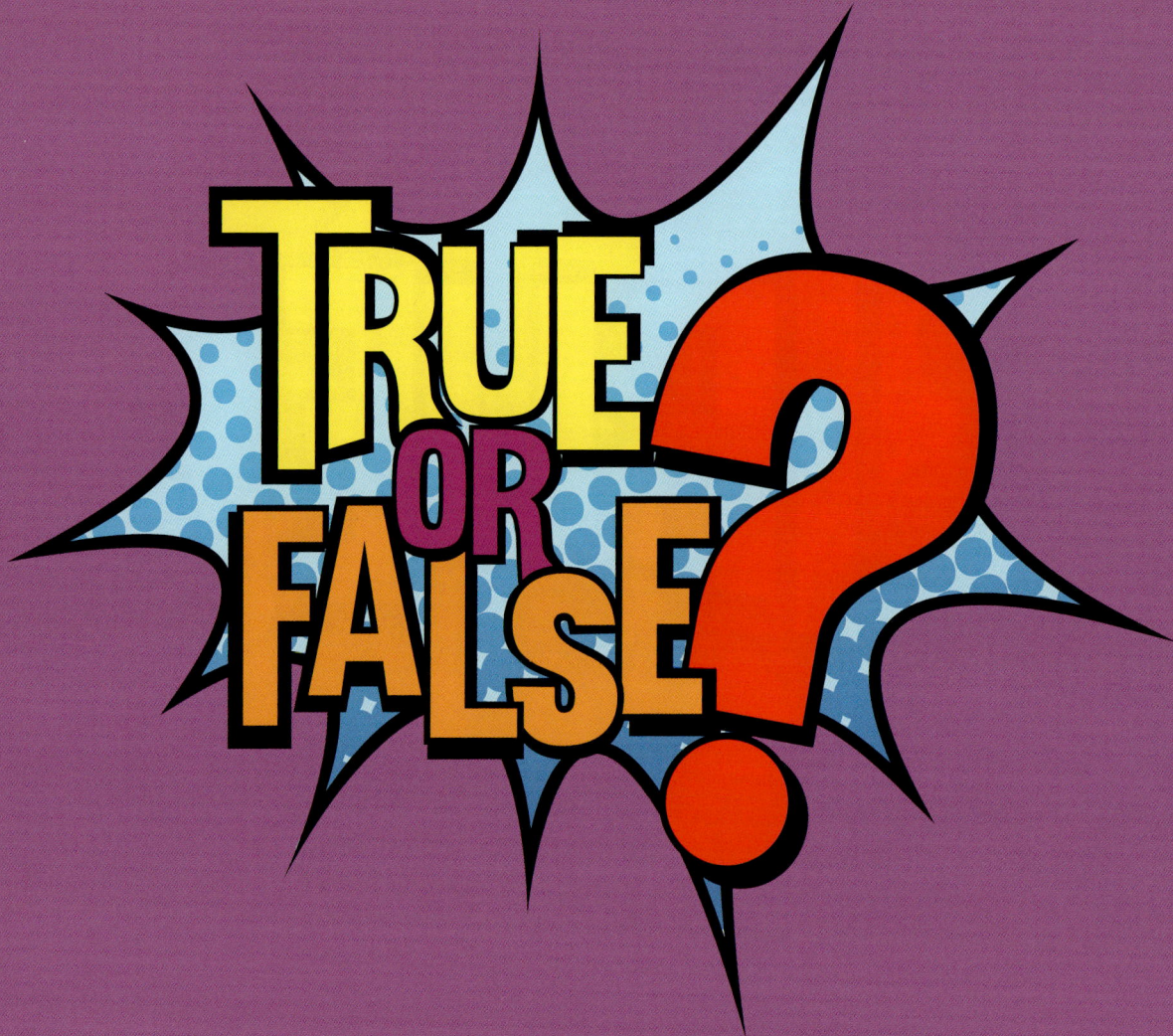

All robots have at least some hard parts.

FALSE!

Engineers are putting more soft, squishy pieces into robots. In 2016, they created Octobot, a robot octopus made of entirely soft parts!

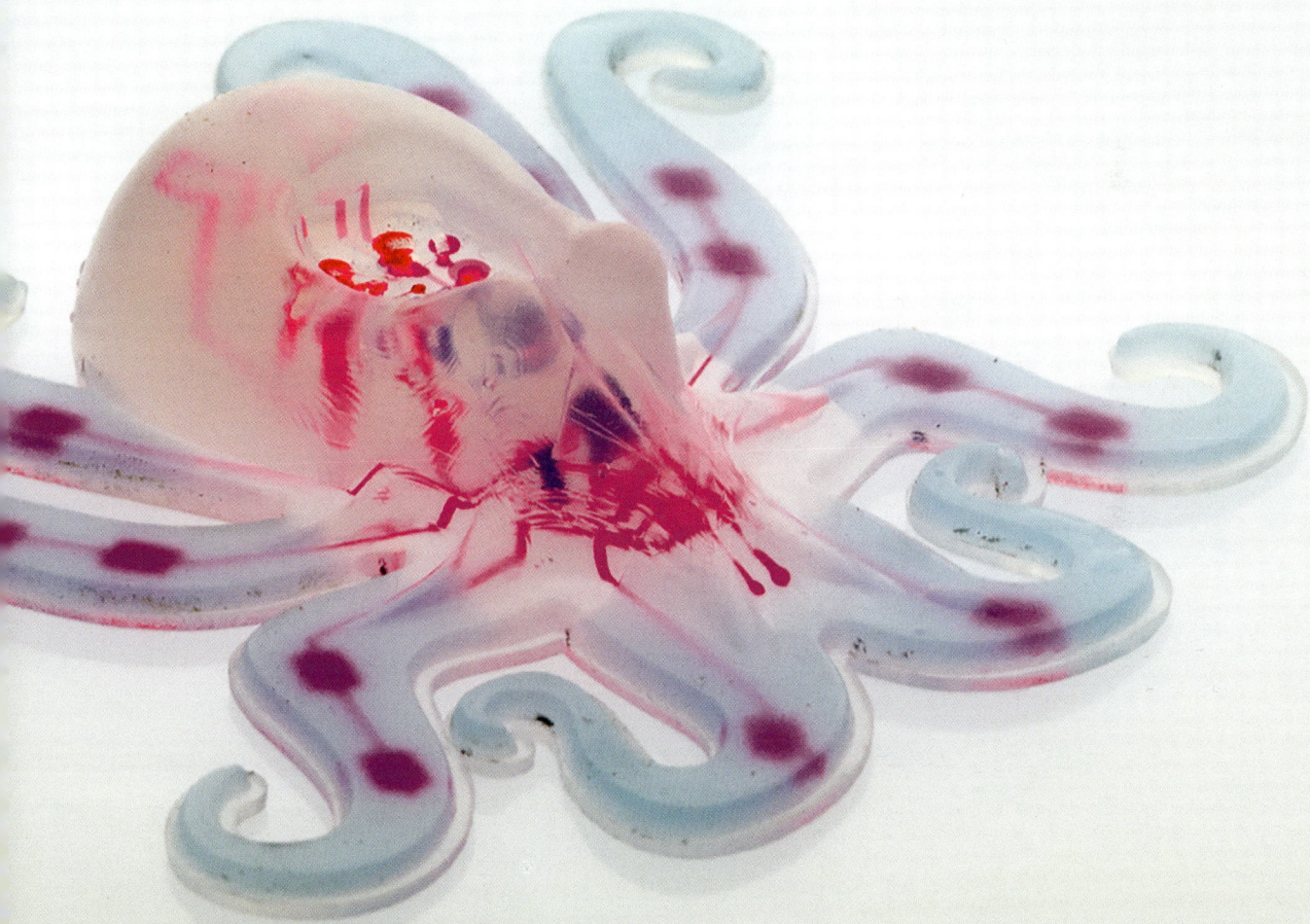

TRUE OR FALSE?

There was a time when most robots worked in the automobile industry.

33

TRUE!

Not too long ago, about 90 percent of working robots were used to build cars! Now, robots do all kinds of jobs, working in hospitals, labs, and even households.

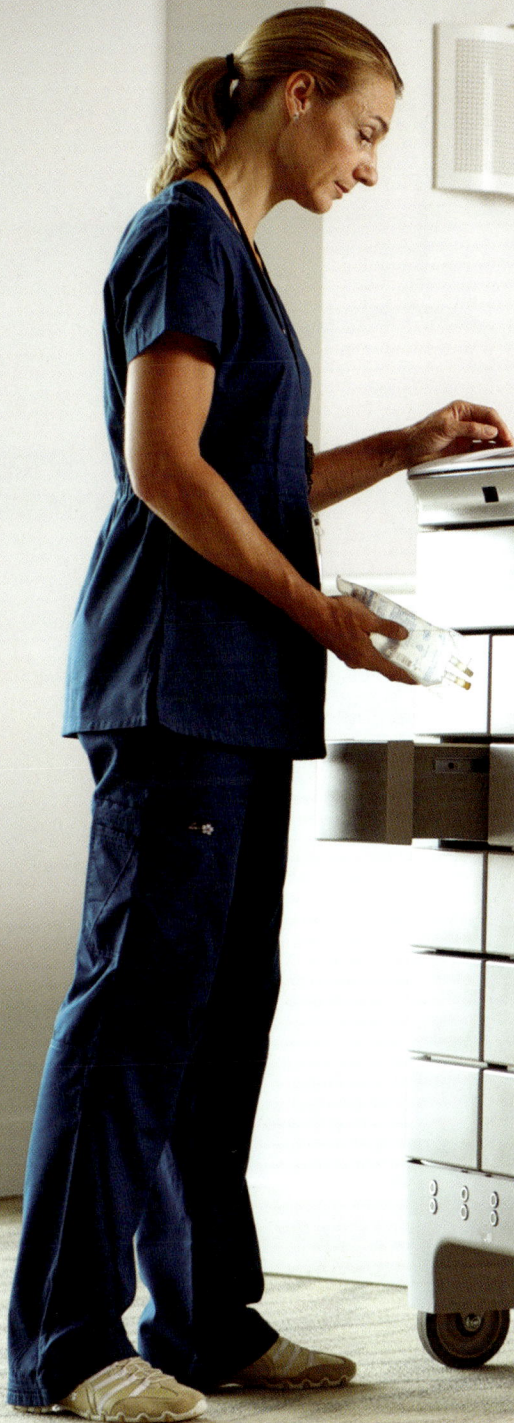

TRUE OR FALSE?

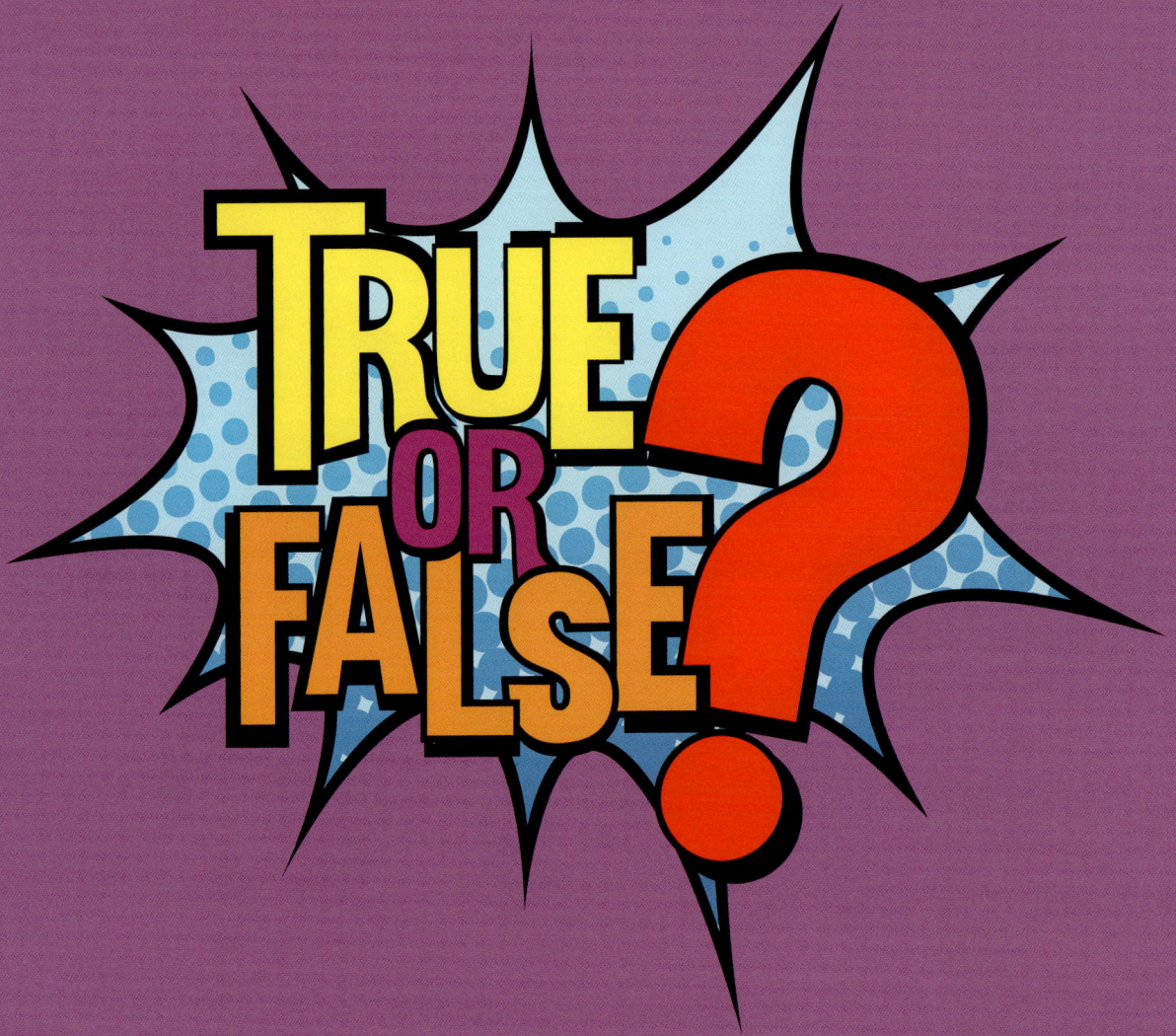

All robots are really, really expensive.

36

FALSE!

Engineers might spend millions of dollars developing a specialized robot. But robots that perform simple, household tasks are available for much less.

38

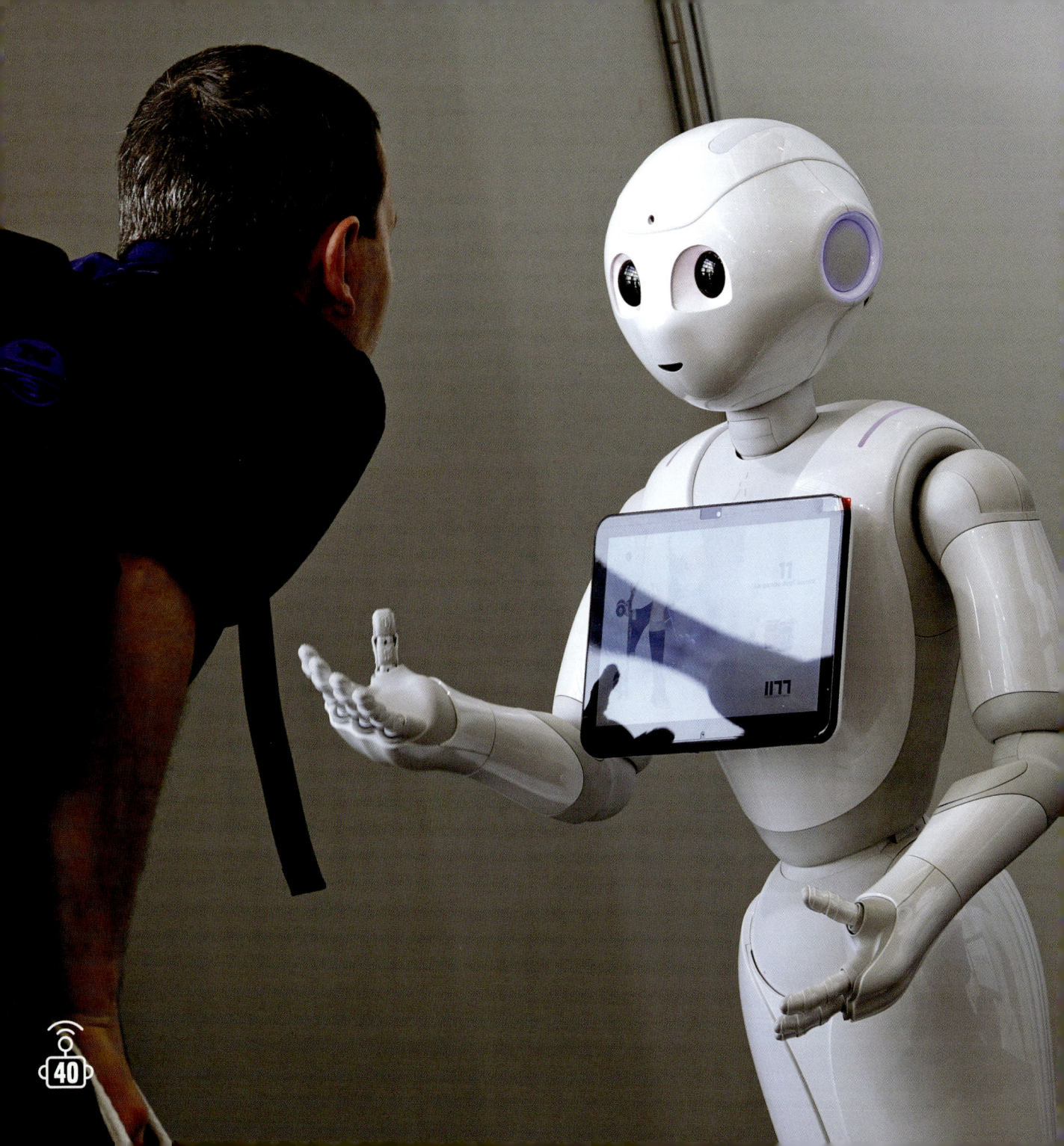

TRUE OR FALSE?

A robot can't be sad, even if its robotic dog runs away from home?

41

TRUE!

Some robots can mimic pain and emotions, but they don't feel these things the way people do.

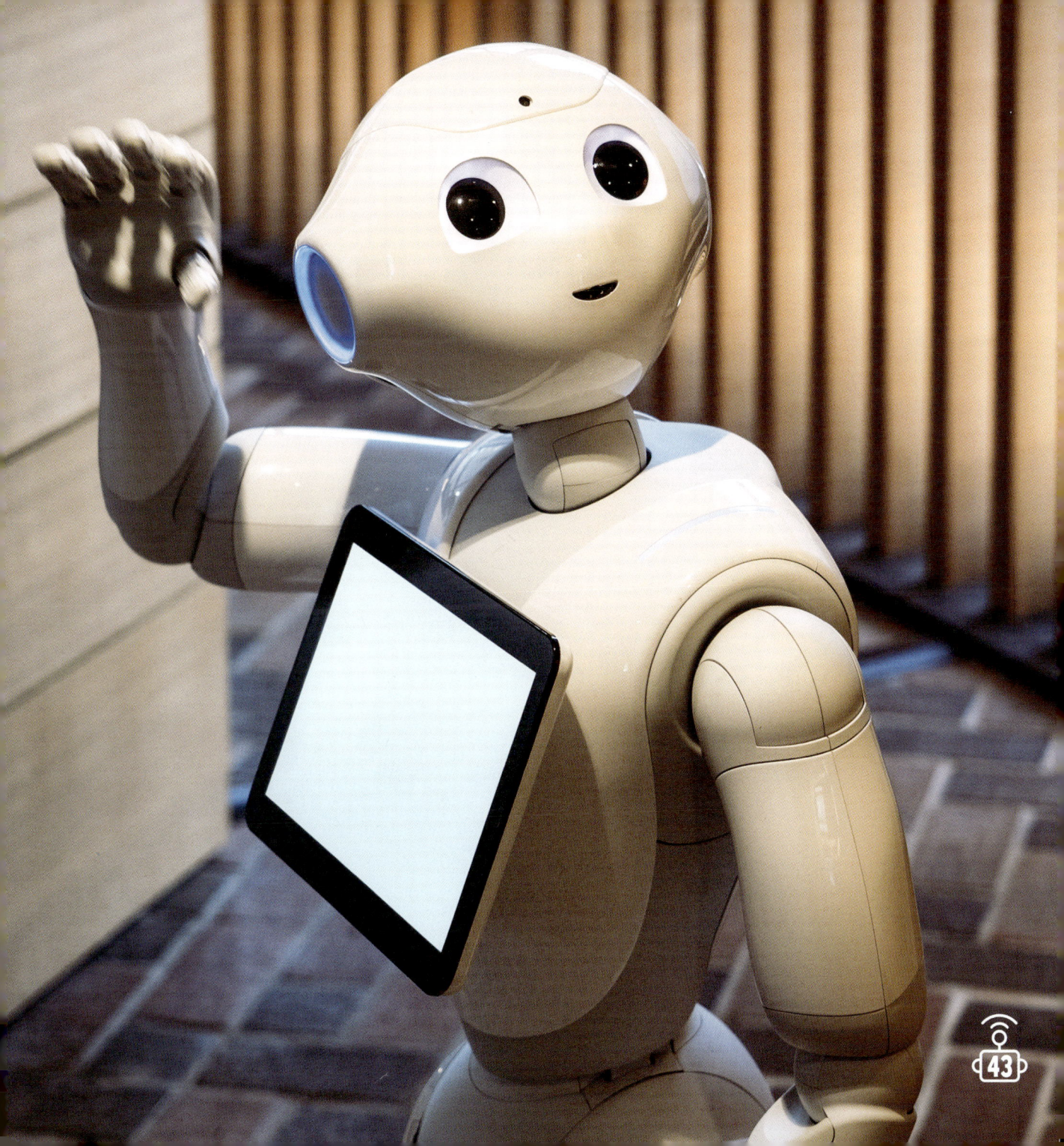

TRUE OR FALSE?

Some robots are dangerous.

45

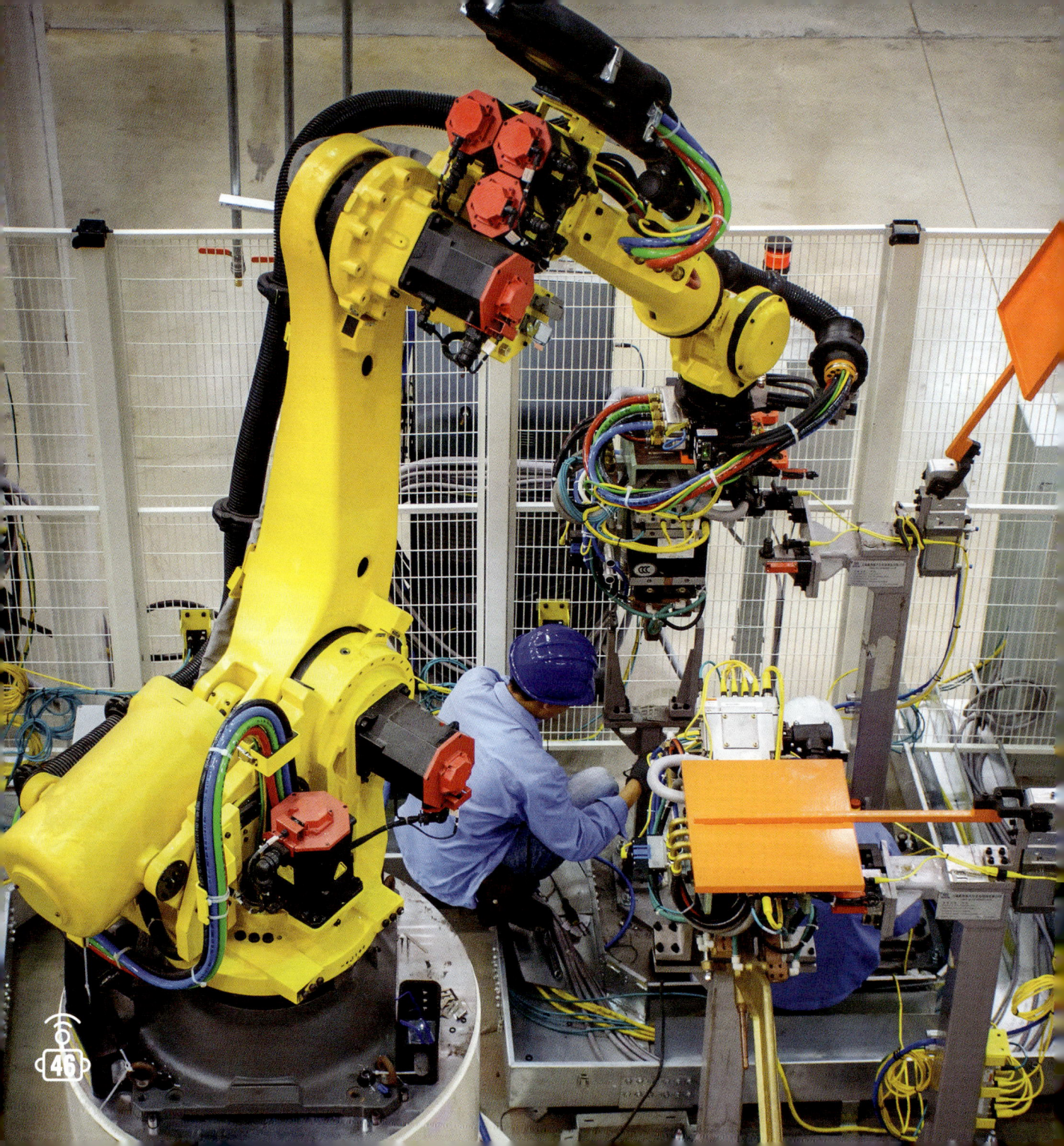

TRUE!

Household robots are low-power and have sensors that prevent them from hurting people. But many industrial robots are quite powerful and have no sensors to tell if people are around. They have to be kept in cages to keep people from getting in their way!

47

TRUE OR FALSE?

Tiny robots can be injected into your bloodstream.

48

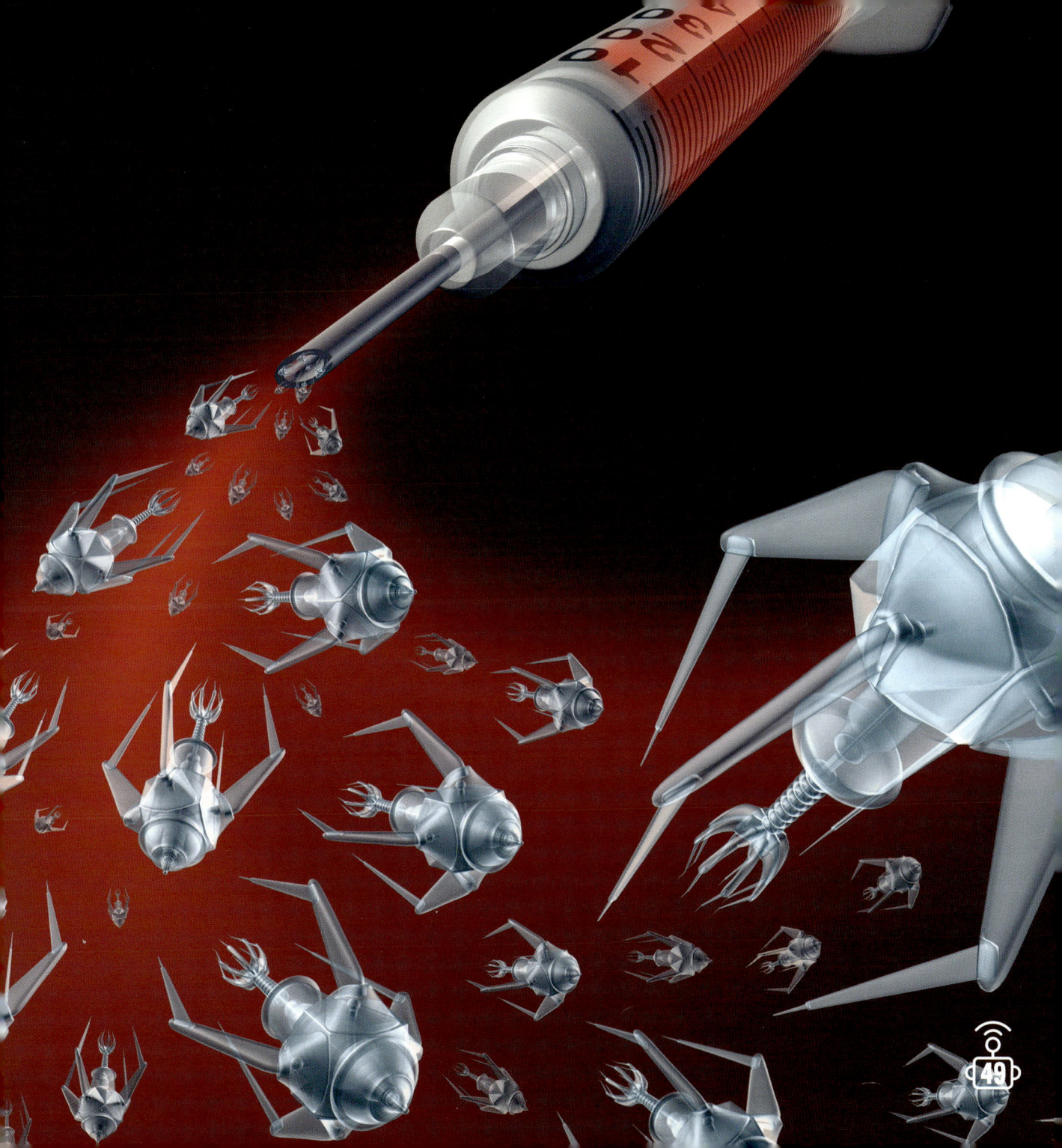

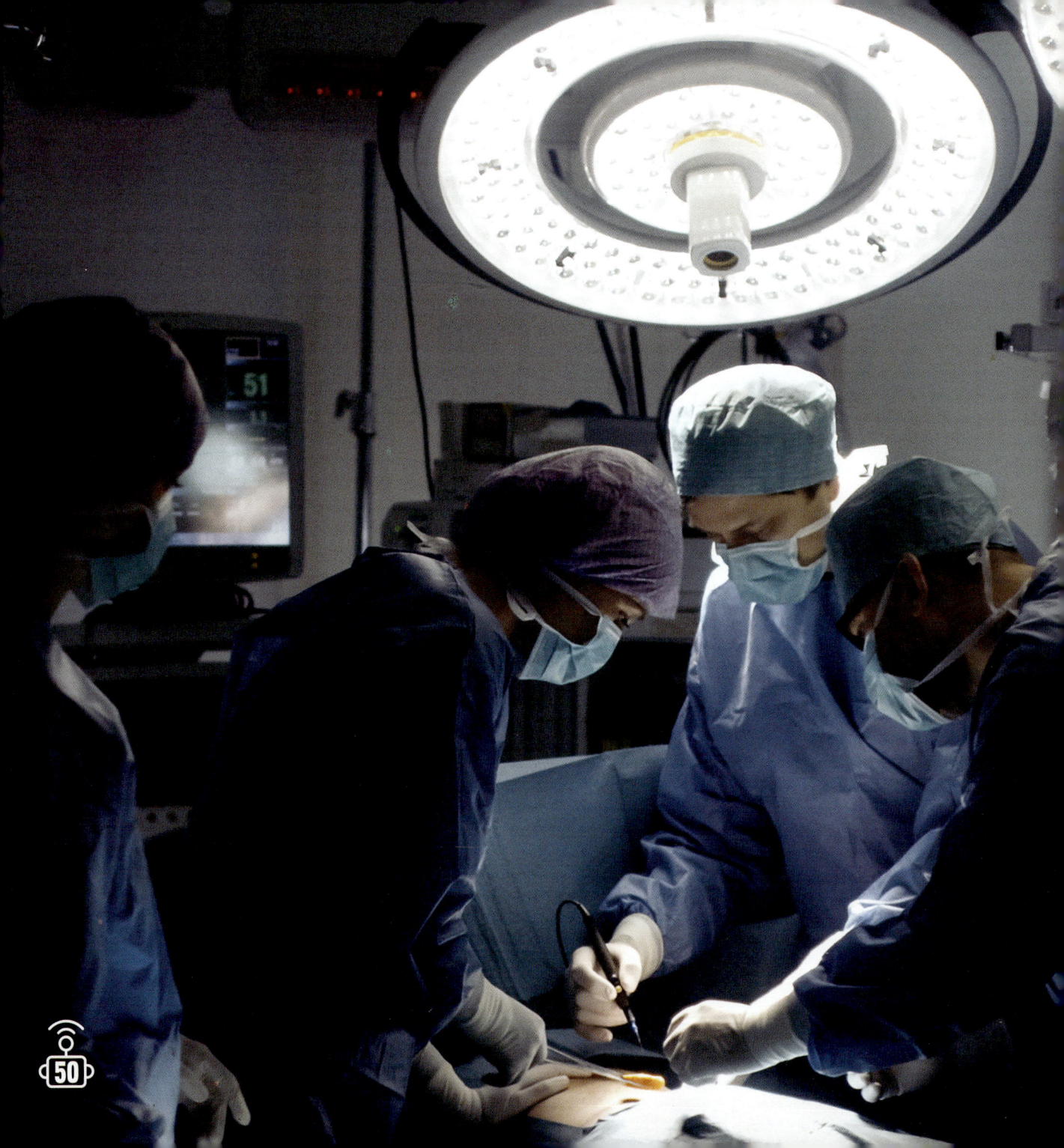

FALSE!

Not yet. But scientists are developing such robots to deliver drugs, attack cancer cells, and perform tiny *microsurgeries.*

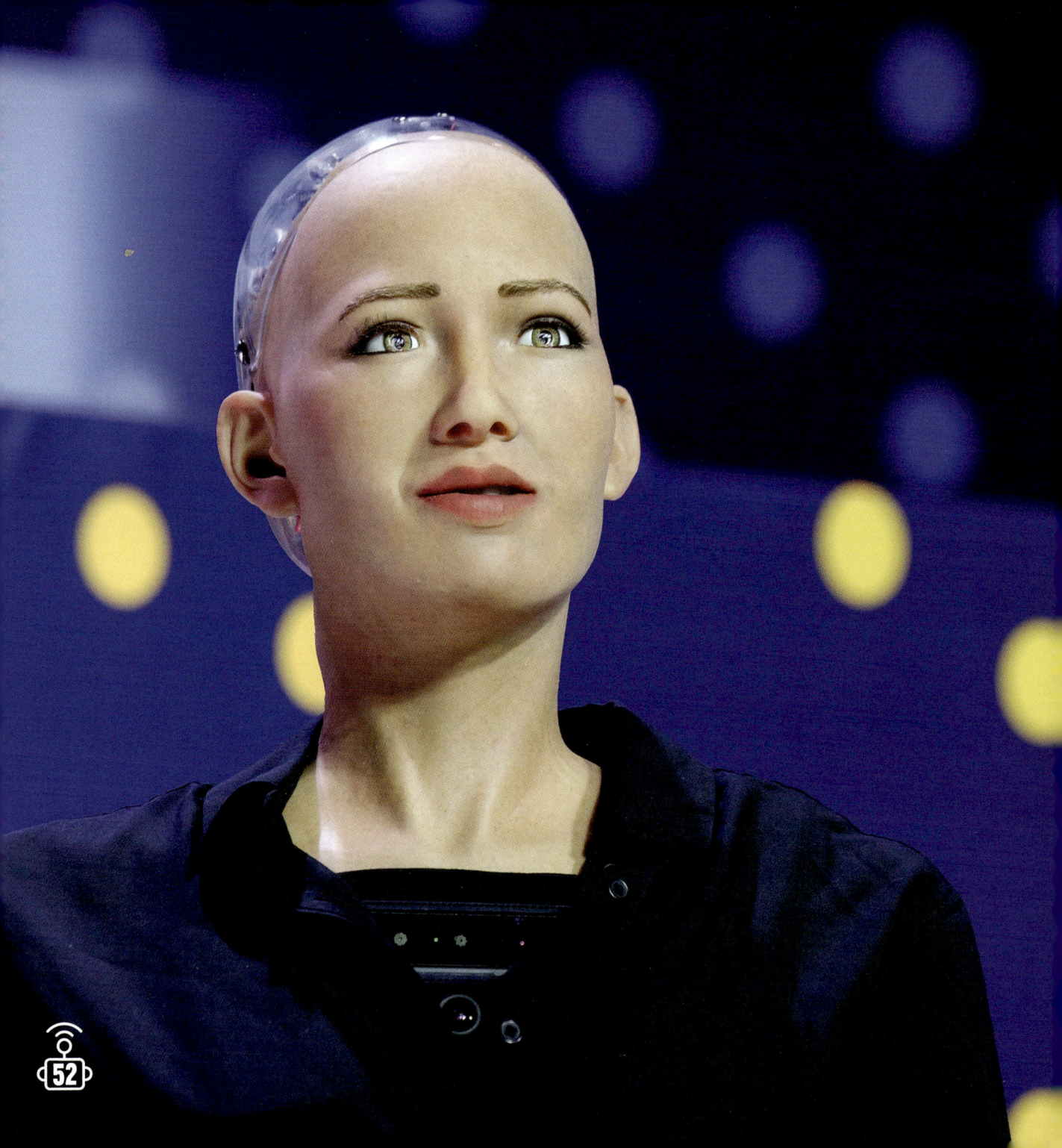

TRUE OR FALSE?

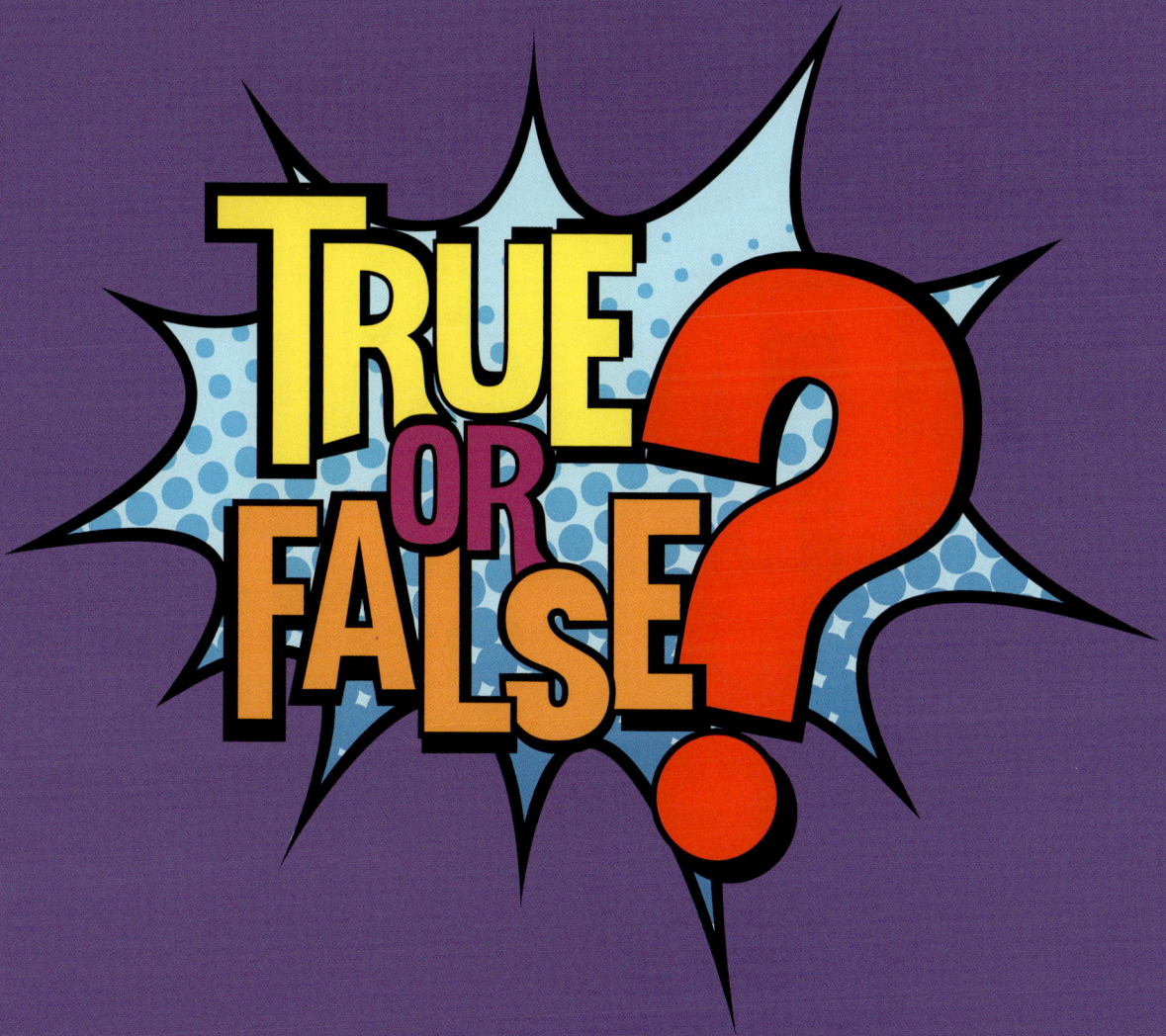

There are robots disguised as people walking among us.

53

FALSE!

...as far as we know. But robots in human disguise are a popular theme in science-fiction stories.

54

TRUE OR FALSE?

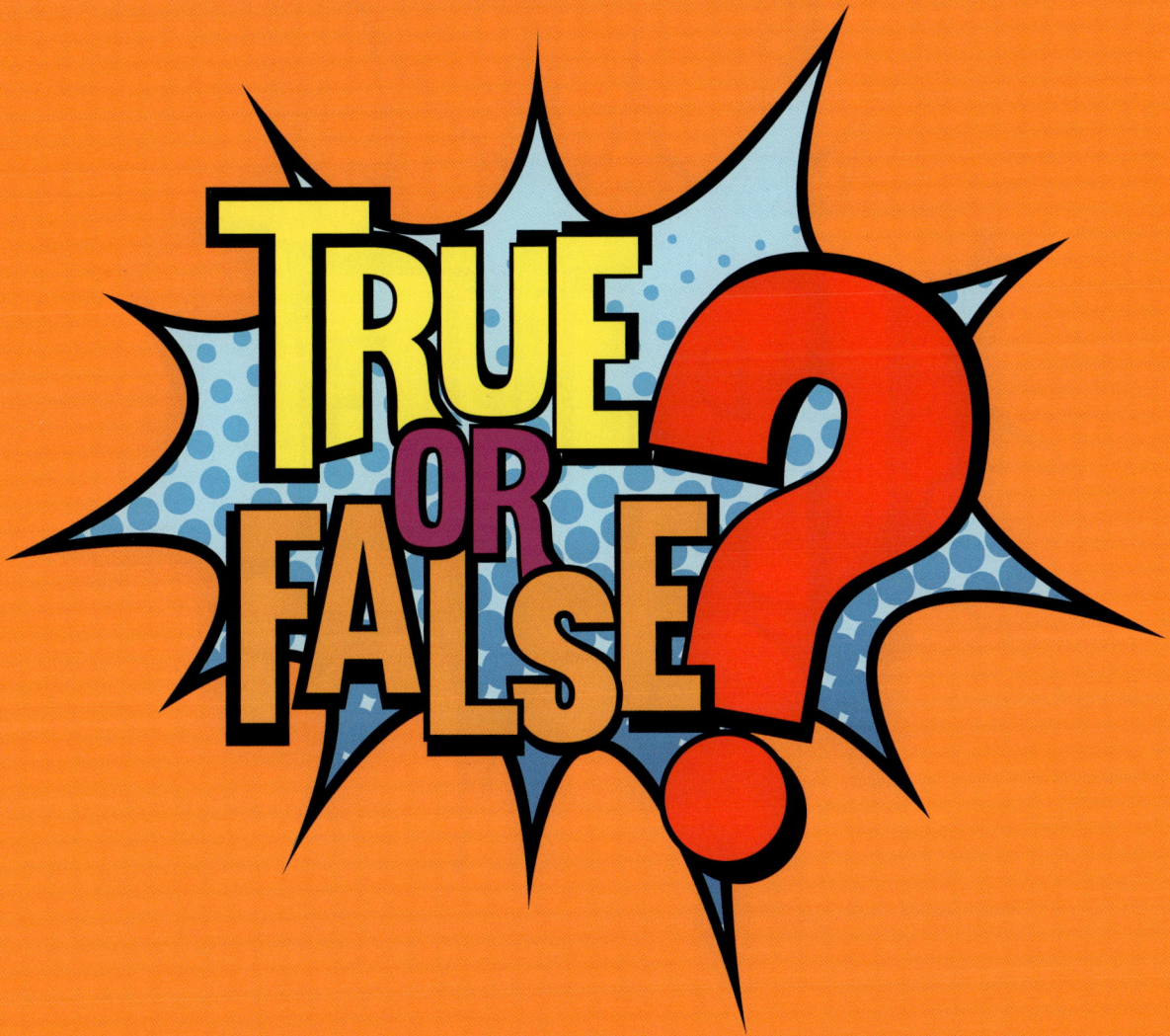

Robots can go places that people cannot.

57

TRUE!

Scientists can use robots to explore destinations that are difficult or dangerous for humans to visit.

58

There are robots on Mars.

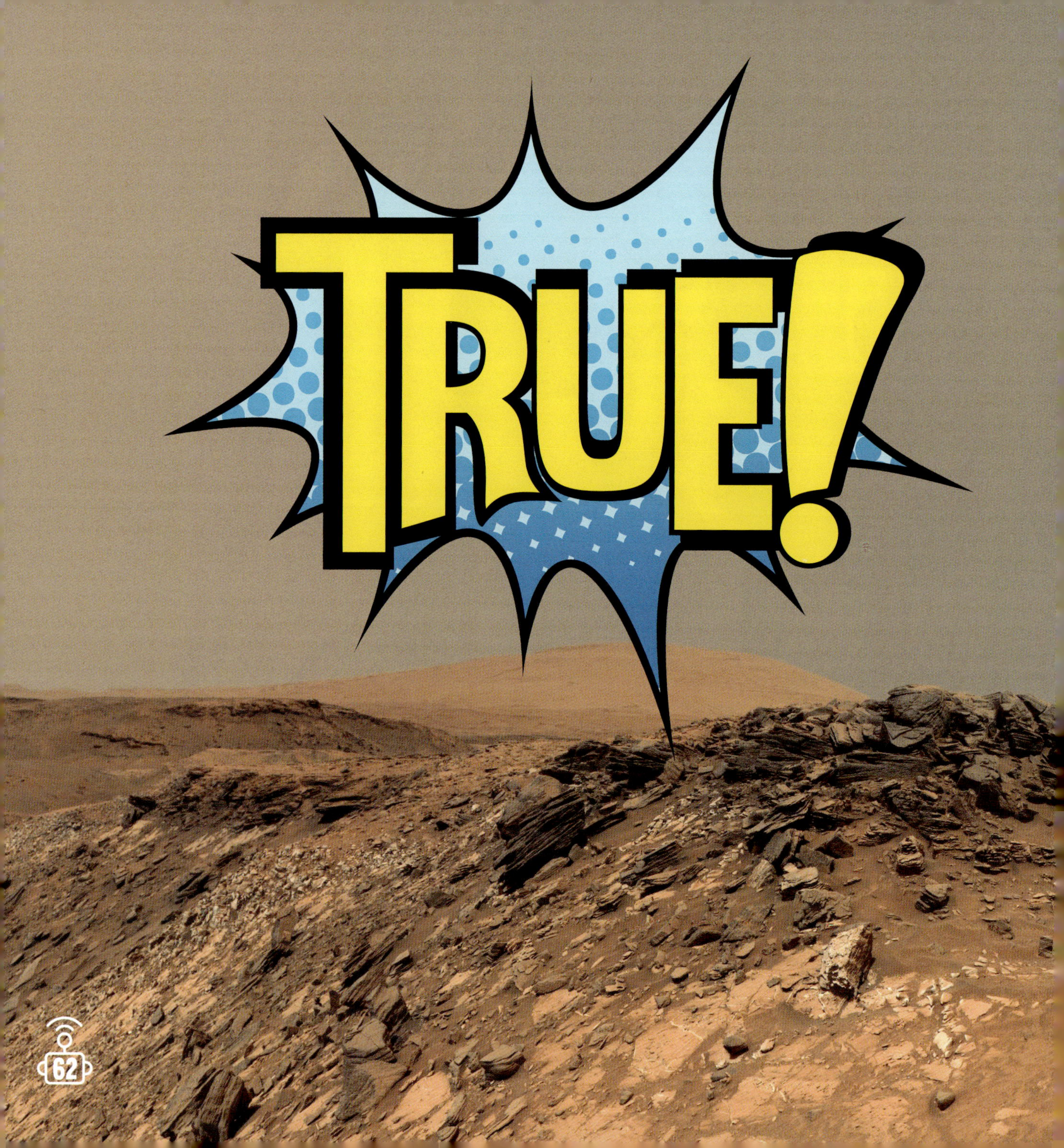
TRUE!

Many robotic space probes have landed on the Martian surface. Curiosity, a roving robot the size of a small car, has been exploring the planet since 2012.

TRUE OR FALSE?

Robot stands for **rapid**, omni-purpose **building** optimization technology.

64

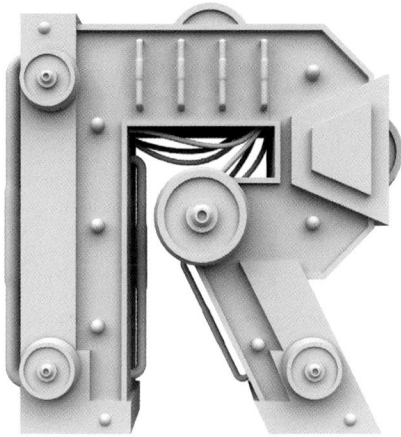

FALSE!

Robot comes from a Czech word meaning *forced labor*. The term was coined for use in a science-fiction play!

67

TRUE OR FALSE?

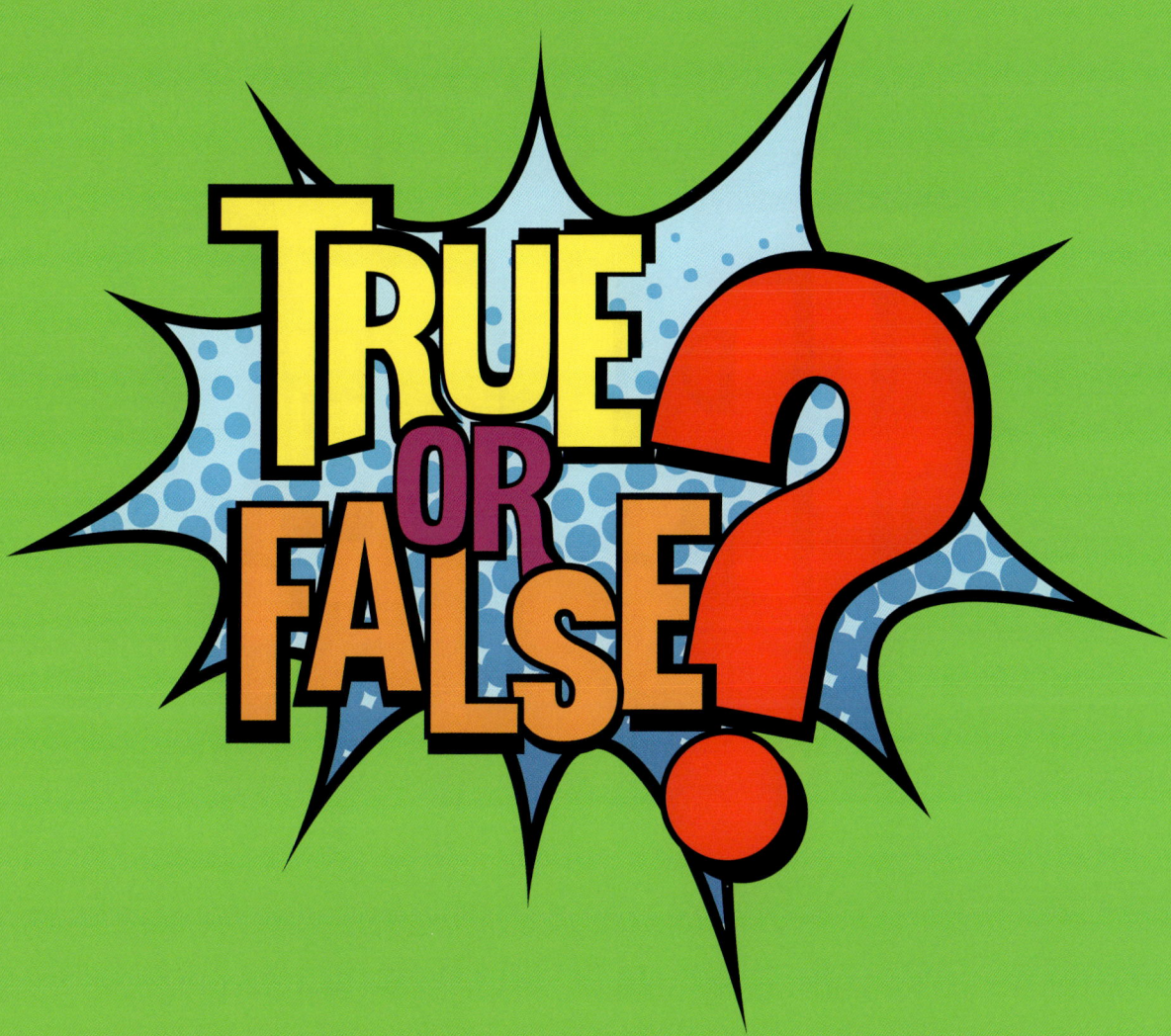

The basic idea of a robot has been around since ancient times.

69

TRUE!

In Greek mythology, for example, the god Hephaestus made a robotlike creation called Talos to guard the island of Crete. But we mortals have only been able to make robots for the last 75 years or so.

70

TRUE OR FALSE?

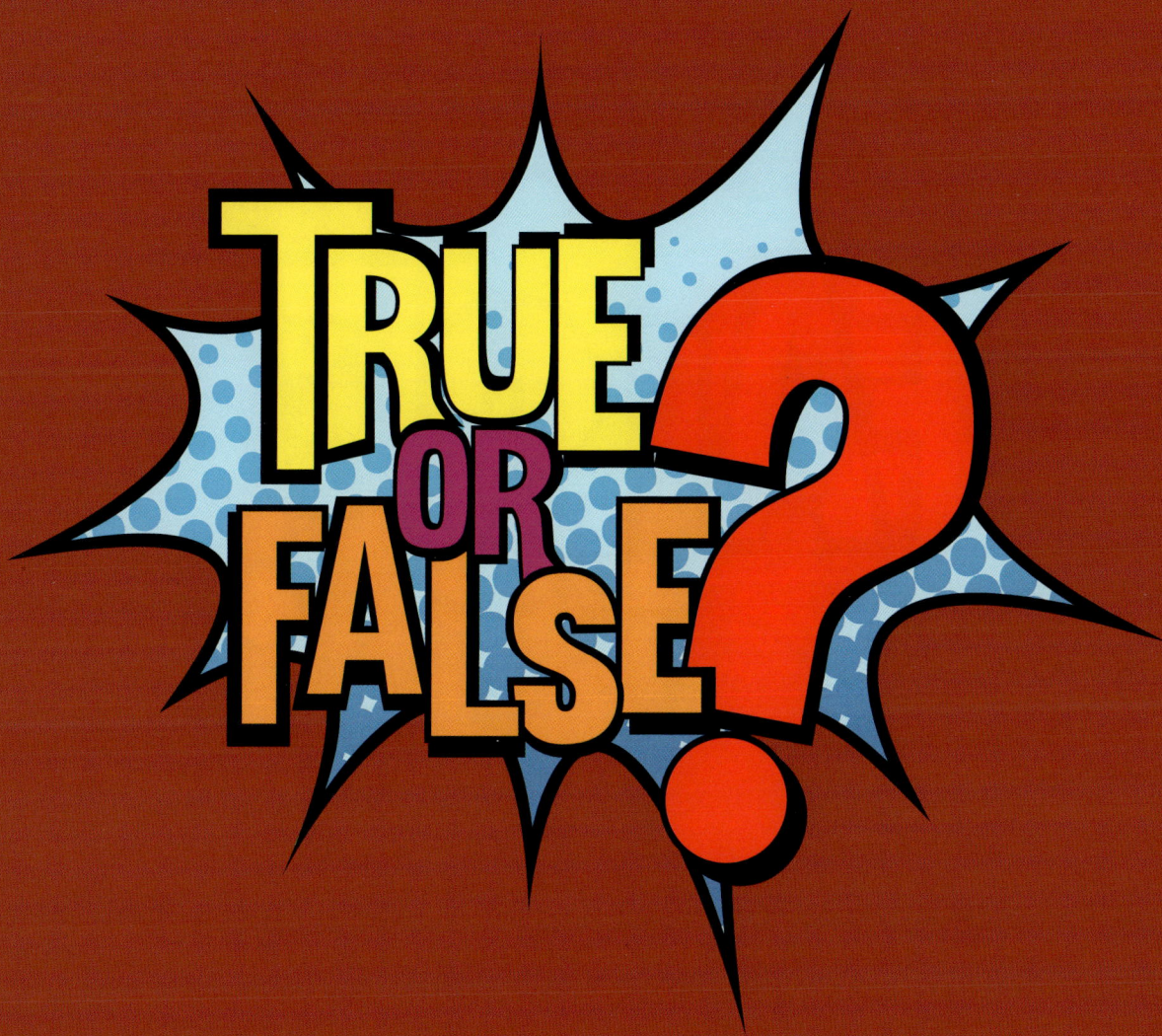

Robots make terrible musicians.

73

FALSE!

Robots can be programmed to reflect the creativity of their programmers. They have been made to play music, paint, and dance. Some can even "create" new works by combining various influences.

TRUE OR FALSE?

Soon, robots might swarm our streets.

TRUE!

The widespread use of self-driving cars may be just around the corner. If all goes according to plan, robotic vehicles will soon get us and our stuff from point A to point B safely and efficiently.

TRUE OR FALSE?

You don't have to say "please" and "thank you" to robots, but it's still the polite thing to do.

TRUE!

Robots that respond to voice commands generally don't need to hear "please" or "thank you." But it's good to stay in the habit of using polite words, so you won't forget them when you ask for help from people!

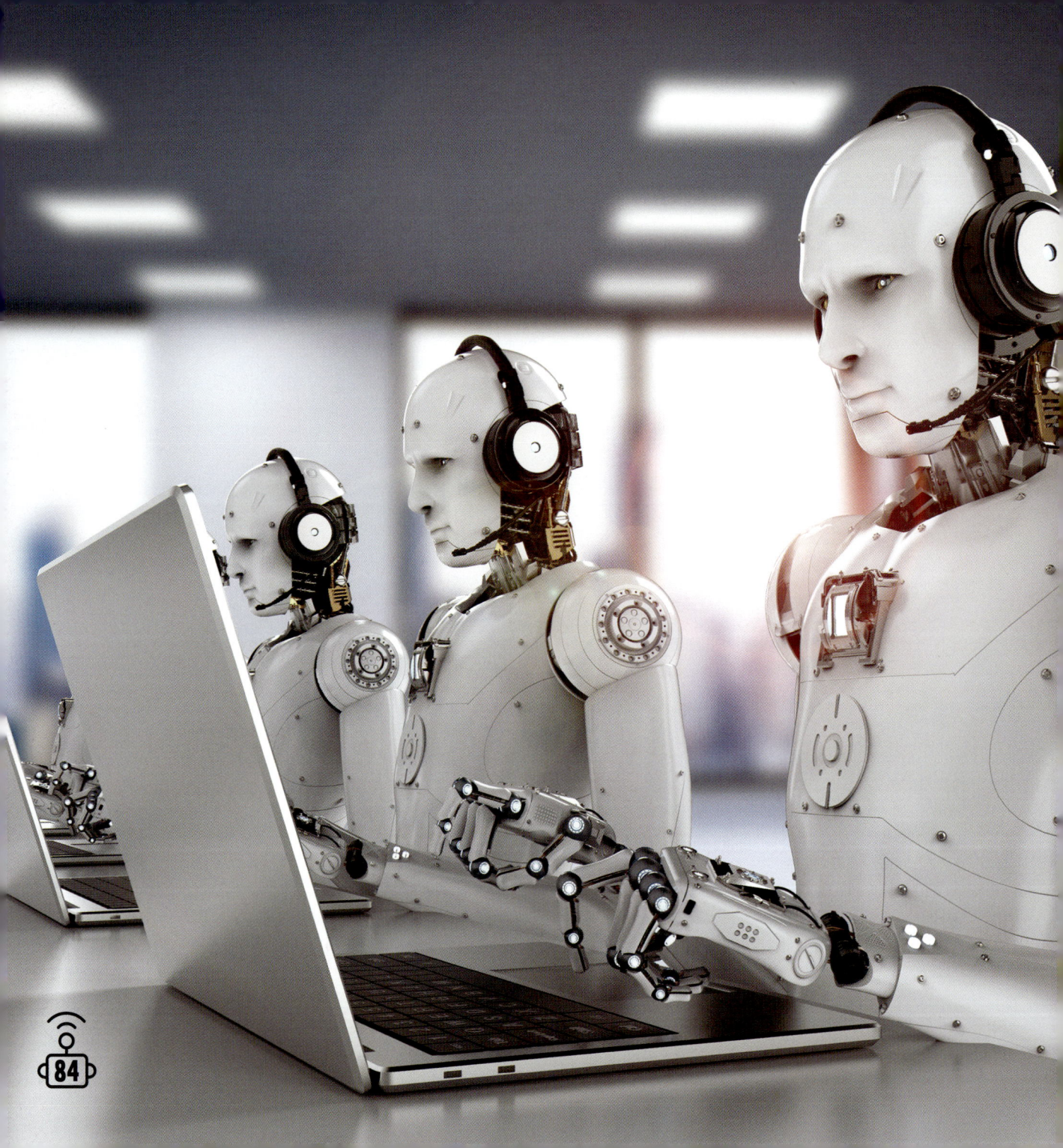

TRUE OR FALSE?

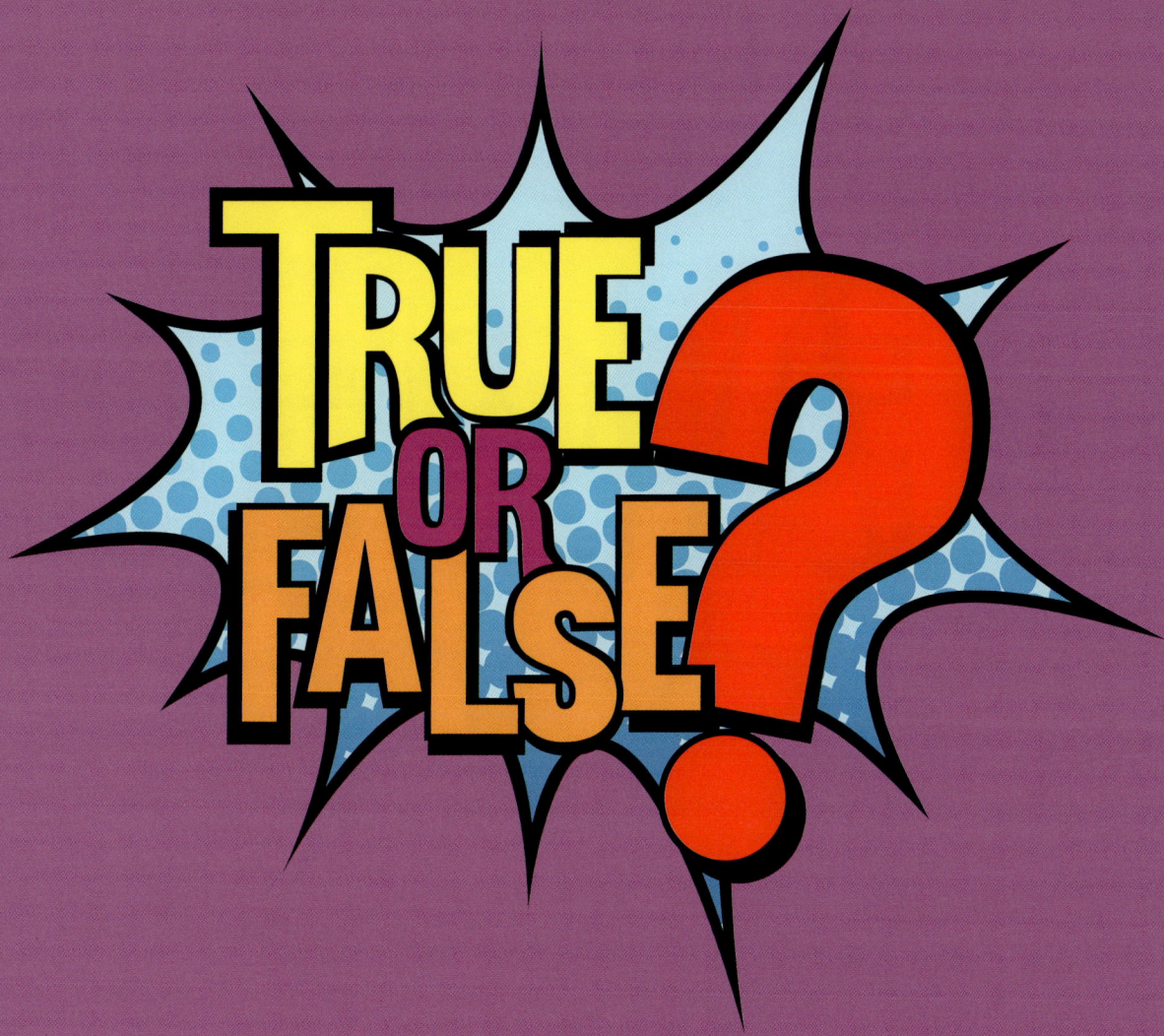

Robots will take over everyone's jobs someday.

85

FALSE!

But, they might take over a lot of jobs. One thing is certain—there will be plenty of jobs designing, building, and repairing robots in the near future.

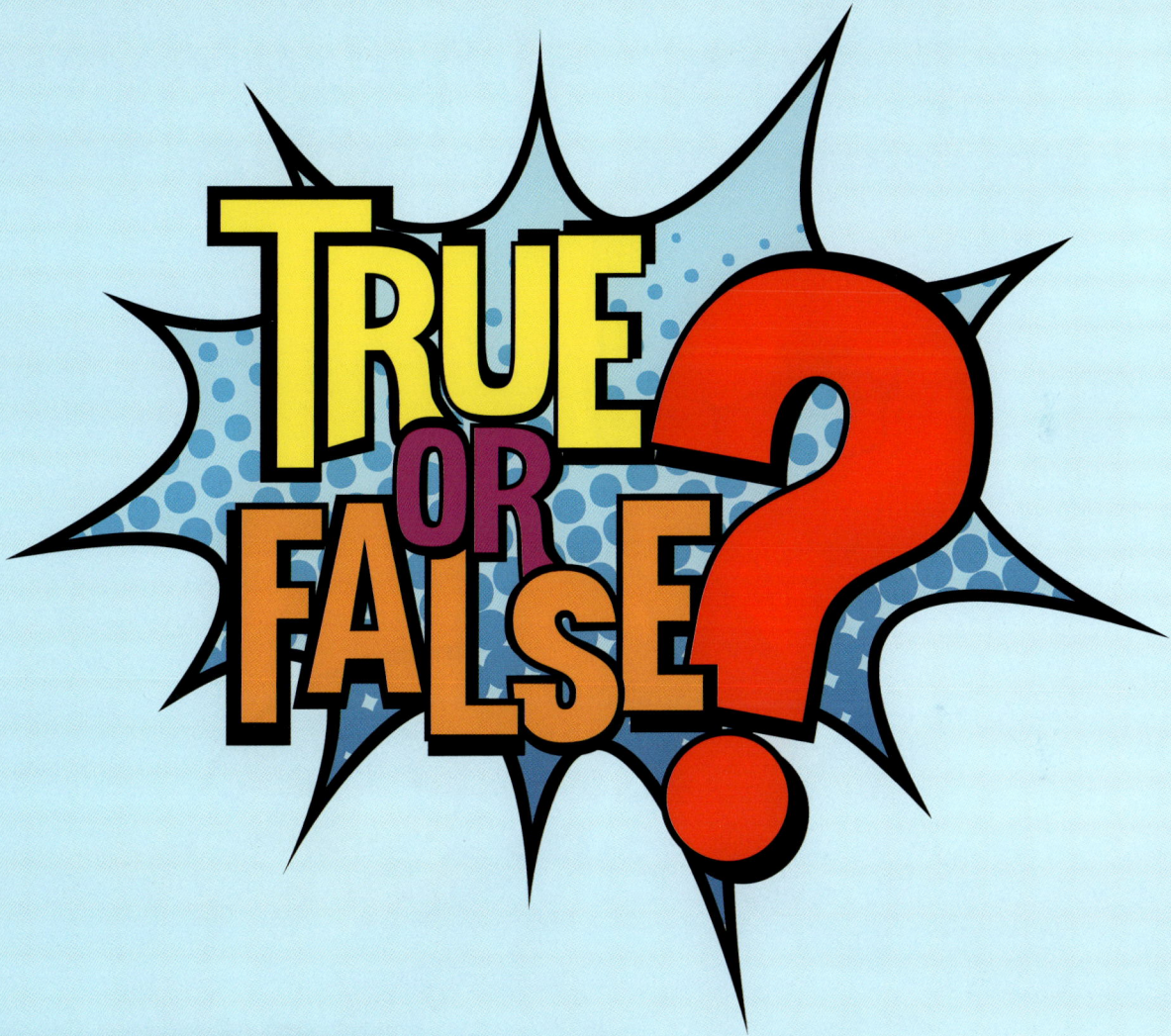

TRUE OR FALSE?

Only scientists and engineers can design and build robots.

FALSE!

You can build a robot. Yes, you! Websites, books, building kits, and robotics clubs can help you get started.

DID YOU KNOW...

Robotic security guards patrol some shopping centers and parking lots. They alert a human guard at the first sign of trouble.

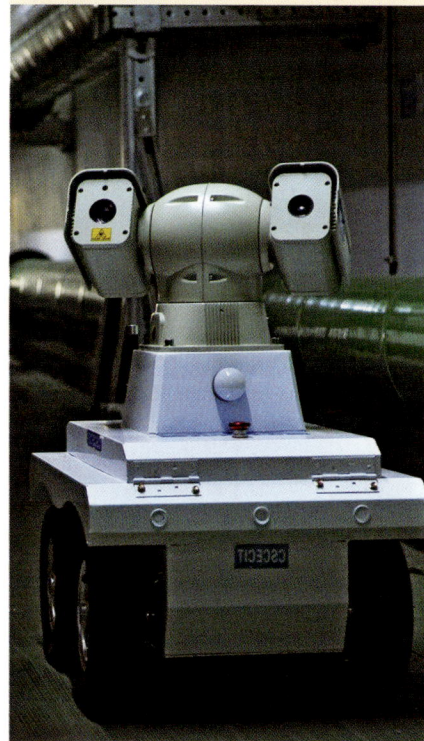

Mobile robots helped firefighters to save much of Paris's famed Notre Dame cathedral during a serious 2019 fire.

A humanlike robot named Sophia was named an honorary citizen of Saudi Arabia.

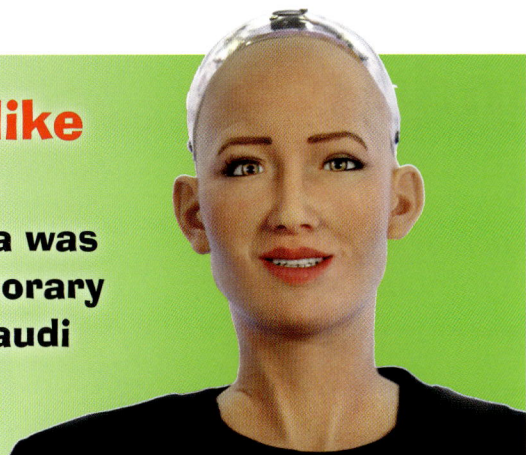

In Japan, there is a huge factory where industrial robots make other industrial robots— **no humans needed.**

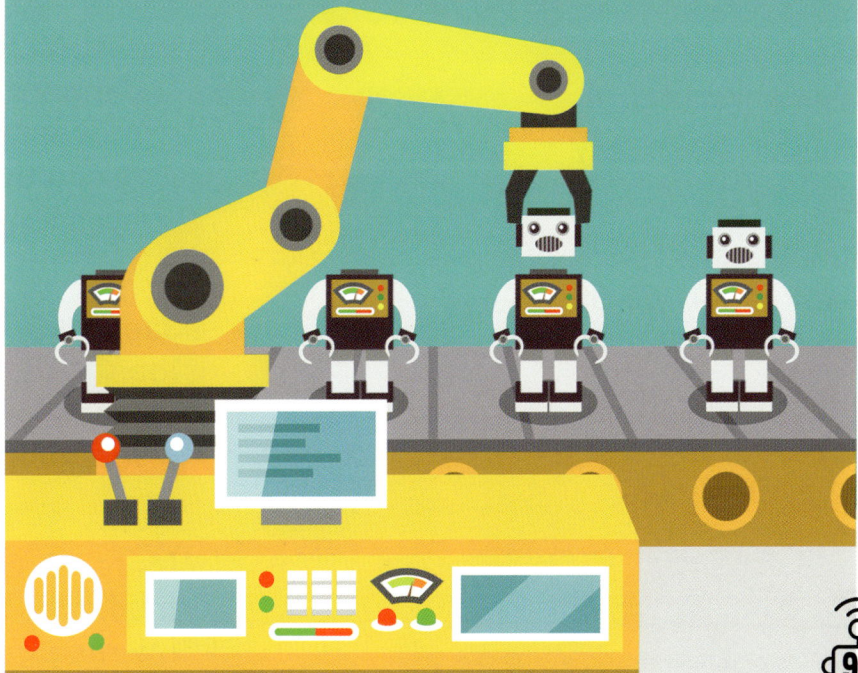

The first true robot was nicknamed **Shakey** because it was tall and wobbled when it moved.

Index

Acknowledgments

Cover: © G Tech/Shutterstock; © BSD/Shutterstock; © Good Ware/Shutterstock

5-11   © Shutterstock
12   © Yoshikazu Tsuno, Getty Images
14-25   © Shutterstock
26   © Adam Berry, Getty Images
29   © Andriy Shevchuk, Shutterstock
30-31   © Lori Sanders, Harvard University
32-33   © Jenson/Shutterstock
35   © Aethon
37-61   © Shutterstock
62-63   NASA/JPL-Caltech/MSSS
65   © Shutterstock
66   © Imagno/Getty Images
68   © Alpha Media Production/Shutterstock
71   Public Domain
72   © Mimi Haddon, Getty Images
75-79   © Shutterstock
81   © George Rinhart, Getty Images
83-90   © Shutterstock
92-93   © Shutterstock; © Visual China Group/Getty Images; © Ralph Crane, The LIFE Picture Collection/Getty Images